21 Things to Do After You Get Your Amateur Radio License

Dan Romanchik, KB6NU

ISBN: 0692399992
ISBN-13: 978-0692399996

CONTENTS

INTRODUCTION

It used to be that the pathway into amateur radio was clear. You got a Novice license, operated CW (Morse Code) on the shortwave bands for a while using a crystal-controlled, relatively low-power transmitter, then got your Technician Class or General Class license. You had to get that next license within a year because that's when your Novice class licensed expired. It was not renewable.

Today, that path is not so clear. The Technician Class license is the license that most newcomers obtain first, and with that license, they can do much more than Novices could do way back when.

Even with all the additional privileges, it sometimes seems that new hams are not as active as Novices were back in the day. Maybe it's because there is so much that they can now do, or maybe because they lack guidance as to what they should be doing to get their feet wet in this great hobby.

If it's guidance that you're seeking, you've come to the right place. This book suggests 21 different things to do after you get your ham radio license. Some of these suggestions will sound like work, and they are work, but they will also help you have more fun with amateur radio. Here are the 21 things you should do after you get your amateur radio license:

1. Join a club
2. Join the ARRL
3. Find an Elmer
4. Buy a radio
5. Get on the air
6. Set up a shack
7. Buy some tools
8. Buy a DMM
9. Build an antenna
10. Build a kit
11. Go to a hamfest
12. Learn the lingo
13. Subscribe to some mailing lists, blogs, and podcasts
14. Upgrade to General
15. Go to Field Day
16. Learn Morse Code
17. Get to know your (amateur radio) neighbors
18. Buy some QSL cards
19. Join SkyWarn, ARES, or RACES
20. Participate in a contest
21. HAVE FUN!

73,

Dan Romanchik, KB6NU

JOIN A CLUB

One of the very first things you should do after you get your first amateur radio license is to join a club. There are many reasons why this is a good idea, but perhaps the biggest reason is that amateur radio is more fun when shared with others. The whole point of amateur radio is to make contacts with other amateur radio operators. By joining a club, you start making face-to-face contacts.

All of the other reasons for joining a club stem from this idea of sharing the hobby with other amateur radio operators. For example, you can think of the other members of the club as a vast reservoir of knowledge that you can tap.

Want some advice on what radio to buy? Ask a club members. Need some help installing an antenna? Ask a club member. Have a question about the best place to buy feedline or connectors? Ask a club member. I think you get the idea.

Clubs conduct a variety of activities that you'll find both interesting and useful. Many clubs, for example, have speakers at their monthly meetings that discuss some aspect of amateur radio. By attending these meetings, you'll not only learn about the topic, but have someone that you can contact should you decide to pursue that topic further.

Clubs also hold classes and administer license

examinations. Being a member of the club will make it easier for you to take advantage of the classes and help you upgrade your license more easily.

Another benefit that some clubs offer is the use of a club station. This station may allow use to use equipment or operate modes that would be impossible to do at your home station. Our club station, for example, has a three-element Yagi antenna up about 70 feet. There's no way that I could install such an antenna system at my home. Using the club station, though, allows me to experience using this antenna system and learn all about how they work and how well they work.

Being a club member can even help you get a good deal on used equipment. Club members often offer their used gear to other club members at a lower price than they would ask if they listed them online or taking them to hamfests. Not only do you get a lower price, but it's less likely that there will be a problem with your purchase, and if there is, you know exactly where to find the seller.

Finding a club

If you don't know of any clubs in your area, go to http://www.arrl.org/find-a-club. Type your zip code into the appropriate box, and soon you'll get a list of clubs in your area.

The listings will show what services the clubs offer, their specialties, and if the club has a website, the website address. This information should give you an idea of how active the club is and what kinds of things the club members are interested in.

If there are several clubs in your area, visit them all before deciding to join one. Just like people, clubs have their own personalities, and you may find that you fit better with one club rather than another. For example, some clubs emphasize emergency communications and public service. If you're not really interested in those activities, that club may not be for you.

Whichever club you choose, go to the meetings and participate in their activities. One thing is certain. You won't get anything out of a club, if you never show up.

JOIN THE ARRL

In addition to joining your local amateur radio club, you should also join the American Radio Relay League (ARRL). The ARRL is that national association for radio amateurs and offers many services for amateur radio operators:

- *QST*. *QST* is the ARRL's monthly magazine. Every month, you'll receive a magazine full of good information, projects, and news about amateur radio.
- *QST* archive. In the past couple of years, the ARRL has digitized every issue of QST that they've ever published. It's available on the ARRL website, but only to current members.
- E-mail newsletters. In addition to QST, the ARRL publishes e-mail newsletters on a variety of topics, including:
    - The *ARRL Letter* is a weekly newsletter with news about amateur radio.
    - *Contest Update* is a bi-weekly newsletter for contest enthusiasts. I strongly suggest signing up for this newsletter, as it's about much more than contesting.
    - The *ARES E-Letter* is for those amateur radio operators involved in public service and

emergency communications.
    - Section and division newsletters from your section manager and division director.
- Technical Information Service (TIS). As a member, you can call or e-mail ARRL Technical Information Service specialists for answers to technical and operating questions.
- Legislative advocacy. The ARRL is the only group that effectively speaks for amateur radio with the FCC and Congress. Without this representation, our spectrum would be fair game whenever the political winds shift.
- Outgoing QSL bureau. QSL bureaus help save postage when sending and receiving QSL cards from foreign stations. Only ARRL members can use the outgoing bureau.
- Equipment insurance. Your homeowner's or renter's insurance may or may not cover your amateur radio equipment. This insurance program covers your amateur station, antennas and mobile equipment should they be damaged by lightning, theft, accident, fire, flood, tornado and other natural disasters.

There are lots of naysayers out there who will advise you to not to join the ARRL and tell you it's a waste of money. I think that you should find out for yourself. Join the ARRL and participate in some of its activities. After you've done that, you can make a better decision about whether membership is right for you or not. For more information on joining the ARRL, go to https://www.arrl.org/join-arrl-renew-membership/.

FIND AN ELMER

Amateur radio can be a complicated hobby. You will, undoubtedly, have questions about the technology, questions about the rules, and questions about operating procedures. An "Elmer" is someone who can help answer those questions and help you avoid some of the pitfalls of the hobby. He or she is a ham that you can go to when you have a question about what rig to buy, when you want to borrow an antenna analyzer, or when you're having trouble understanding a particular concept. If you haven't already, you might want to find an Elmer.

The term Elmer first appeared in the March 1971 issue of QST magazine. In that issue, Rod Newkirk, W9BRD, called them "the unsung fathers of Amateur Radio." He wrote that an Elmer is "the ham who took the most time and trouble to give you a push toward your license."

Where do you find an Elmer? Well, the first place you might look is the club you just joined. Lots of the "old timers" there are more than happy to help newcomers, and many clubs have "Elmer" programs. Ask for help and ye just may receive.

Nowadays, you might find your Elmer online. There are lots of websites and mailing lists that are geared towards helping people become better amateur radio operators. One mailing list that I am a member of is the HamRadioHelpGroup mailing list (http://groups.yahoo.com/group/HamRadioHelpGroup/). In addition to newcomers looking for help, there are experienced guys like me who are happy to help.

You can get by without an Elmer, but without one, it's easy to become frustrated and set aside the hobby. One ham I spoke with said, "I did not have an Elmer. I got my license, and within a year, I started a 20 year hiatus. I blame that on not having an Elmer."

You don't want that to happen to you. Find an Elmer and take his or her advice. You'll get a lot more out of the hobby.

BUY A RADIO

We all remember our first radio. I got my license back in the day when separate transmitters and receivers were more common than transceivers are today. So, my first radio was a combination of a Hammarlund HQ-101 receiver and a Heathkit DX-60B transmitter. With this combination, I was able to operate CW and AM on the 80m, 40m, 20m, 15m, and 10m bands. Of course, because I was only a Novice, I wasn't allowed on 20m at all, and could only operate CW on the other bands.

Today, it's more common for one's first radio to be a VHF or VHF/UHF FM transceiver. Now that some Chinese companies have entered the amateur radio market, you can buy a handheld VHF/UHF transceiver, like the Baofeng UV-5R, shown below, for less than $75 in the U.S. A handheld transceiver makes a good first radio, but remember that it is just your *first* radio. If you never buy a second (or a third or a fourth), you won't be able to take advantage of what our great hobby offers.

The ARRL has actually done a much better job of advising new hams about buying a first radio than I can do here. So, let me point you to the ARRL Web page, “Buying Your First Radio” (http://www.arrl.org/buying-your-first-radio). On this page, there are links to several PDF files that will help you choose your first radio.

The first publication you’ll see there is the 24-page brochure, *Choosing a Ham Radio*. Reading this publication will get you thinking about what kind of operating you want to do, which is really the first step in choosing the right radio for you. Next, it describes features found in modern radios, and you’ll gain a good understanding of how that affects your choice of radio.

Choosing a Ham Radio also contains a lot of information about HF, or shortwave, radios. While you may start out on the VHF and UHF bands, I would encourage you to think about getting on HF, even before you upgrade to General. For me, anyway, the magic of radio is on the shortwave bands, and at the very least, you owe it to yourself to try it.

Buying Used Gear

One approach to getting your first radio is to buy used gear. In general, I would advise against doing this for your first radio. One reason not to buy a used radio is that you're often just buying someone else's problems, especially if you're not in a position to evaluate the condition of a radio. Another reason is that an older radio will not have all the features and could be more difficult to operate than a newer radio.

Having said that, used equipment is not always a bad deal. You might, for example, be able to purchase an older radio from someone you trust, like your Elmer (see Chapter 1) or a fellow club member. When you purchase a radio from someone you trust, not only are you more certain that it will work properly, but you'll have someone to go to with questions or to consult with if there are problems.

Not only that, if you ask nicely, the ham might even let you use the radio for a while before actually purchasing it. I know that I've lent equipment to new hams in the past. Sometimes they decide to buy the radio. Other times, they've decided to purchase a new radio. In either case, they were able to make their decision based on experiences they had with an actual radio.

Finally, don't worry about making the perfect choice. First off, there's no perfect choice, and second, you can always sell the radio and buy something else. Chances are you'll be able to sell it for not too much less than what you paid for it, and you'll have gained a whole lot of experience.

GET ON THE AIR

Once you've acquired a radio, the next step is to get on the air. You're now an amateur radio ***operator***, not an amateur radio listener.

If your first radio is a VHF transceiver, the first thing to do is to find the repeaters in your area. One way to do this is to use the K1IW Amateur Repeater and Broadcast Transmitters Database (http://rptr.amateur-radio.net/). Simply fill in the form, and it will give you a list of repeaters near you. When I asked for 2m repeaters within 10 miles of Ann Arbor, MI, I got the following list:

Location	Frequency	PL (Hz)	Callsign	Sponsor
Ann Arbor, MI	145.23	100.0	W8UM	U of M ARC
Ann Arbor, MI	146.96	100.0	W8PGW	ARROW ARC
Ypsilanti, MI	146.92	100.0	K8RUR	I-94 ARC

What this chart is telling me is that to communicate with other amateur radio operators using the W8UM repeater, I need to set the frequency of my transceiver to 145.23 MHz and enable the CTCSS tone and set that frequency to 100.0

Hz.

Once you've set your transceiver up properly, you should first listen to see if other amateurs are currently using the repeater. If no one is using the repeater, you can try to start a contact with someone, by simply saying your callsign followed by "listening" or "monitoring." For example, I would say, "KB6NU listening." If another amateur radio operator is listening, and wants to talk to me, he or she will replay with his or her callsign, and then we'll start our contact.

If there are already operators using the repeater, listen to their conversation, noting how long they speak before letting the other operator take a turn, how often they identify, and even what they are talking about. If you don't think that they would mind if you joined their conversation, say your callsign immediately after one of them has stopped transmitting. They'll break for you and let you join the conversation.

It really is just that easy. Most hams are welcoming and great people to talk to. By getting on the air frequently, you'll get to know a lot of great people and have many interesting conversations. Sure, you'll run into the occasional grouch, but don't let them ruin your enjoyment of amateur radio.

Operating on the shortwave, of HF, bands is a little different than operating on the VHF bands. A complete description of those differences is beyond the scope of this book, but you'll find a lot of good information in the *The ARRL Operating Manual for Radio Amateurs* (see below).

The basic principles apply, though. Listen first to see if there are other amateur radio operators on the air that you can contact, and if not, give out a call inviting other operators to contact you. Then, once you have established contact, have an interesting contact with the other radio amateur.

Over the years, I've met many amateurs who studied hard to get their license, but then didn't get on the air for months, or even years. Don't let that happen to you. Amateur radio is a contact sport. Get a radio and get on the air!

Resources

- *The ARRL Operating Manual for Radio Amateurs* (http://www.amazon.com/ARRL-Operating-Manual-Radio-Amateurs/dp/0872591093/).
- *FM101x: Using FM Repeaters.* (http://www.amazon.com/Fm-101X-Using-Repeaters-Ac6Vs/dp/143489147X) This book by AC6V (SK) is intended for newcomers to amateur radio.
- A New Ham's Guide: How to Use Amateur (Ham Radio) Repeaters - http://www.hamuniverse.com/repeater.html.

SET UP A SHACK

Setting up a "shack" is an essential part of the ham radio experience. For most amateurs, their shack is a combination workshop and operating position. Another way to think about it is the place where you can go to get away from the hustle and bustle of every day life and immerse yourself in your ham radio hobby.

How did we come to use the term "radio shack"? Well, according to Rod, AC6V, the first radio shacks were found aboard ships in the early 1900s. He says, "At the time, wireless equipment aboard ships was generally housed above the bridge in a wooden structure that was called the 'radio shack'". For many commercial stations, the radio equipment was housed in a shack at the base of the antennas.

An early radio shack can be seen below. This is the shack of amateur radio station 8BNY circa 1922. As you can see, there's not much in the way of amenities.

This 1922 photo shows the "shack of amateur radio station 8BNY.

Setting up your own shack

When you set up your own shack, there's no need to be as ascetic as our forefathers shown above. In fact, I'd advise you to make your shack as comfortable and as convenient as possible. The reason for this is that the more comfortable and convenient it is for you, the more you'll enjoy it, and the more you'll want to use it.

The first thing to think about is where in your house, condo, or apartment you will be setting up your shack. Lucky hams have a spare room that they can use for their shacks. Unfortunately, most of us aren't so lucky. Mine, for example is in my basement. The good thing about that location is that I have plenty of space. The bad thing is that it can get quite cold down there in the winter. All locations are going to have plusses and minuses, so weigh them carefully before getting started.

Richard, K8JHR, is a recently-licensed ham who has given a lot of thought to how to set up a shack. He recently wrote about his experiences and posted the article to the HamRadioHelpGroup, a Yahoo Group for hams looking for help and willing to help. I thought so much of the article, that I reprinted it on my blog at http://www.kb6nu.com/

building-a-new-shack/. You can read the entire article there, but here are some of the more important points:

- Have plenty of AC outlets.
- Buy a big desk that can accommodate lots of radios and station accessories.
- Plan ahead for routing LOTS of wires, cables, and connectors. Think about how you're going to run these cables into and out of your house.
- Buy a really good, substantial, large swivel desk chair.
- Locate your shack as close to the ground as possible.
- Build shelf-risers that give you more vertical space.
- Include space for bookshelves and maybe a filing cabinet.
- Make sure your shack is well-lit.
- Buy a big clock that shows Zulu or UTC time.
- Get a bulletin board for displaying cool QSL cards, certificates, and for posting the odd note.

Another thing to consider when setting up your shack is how to ground your station. Ben, N2IHK, suggests reading W8JI's Web page on ground systems (http://www.w8ji.com/ground_systems.htm). Remember, a good RF ground is not the same as the AC power ground!

Another good reference on setting up an amateur radio station is *The ARRL Handbook for Radio Communications*. This book includes a chapter on station layout and accessories. I've found that the earlier handbooks are better in this regard than the latest editions. For example, my 1986 edition of the *Handbook* includes diagrams on how to build an operating desk, including vertical risers. The 2005 edition does not have those plans.

A good shack will make you a better amateur radio operator. You should give it as much thought, or even more, than you give to buying your first radio.

BUY SOME TOOLS

Like any pursuit, to do the job right, you need to have the proper tools. Amateur radio is no exception. To do certain things, you'll need tools that you may not currently have. Without them, you'll seriously handcuff yourself when it comes to enjoying amateur radio.

You may already have a set of hand tools. Most homeowners, for example, have a hammer, a set of screwdrivers, a set of wrenches, and some pliers to make common home repairs. All of these tools will be useful for amateur radio work, but you'll also need some tools specifically designed for working with electronics, including:

- Needle-nose pliers. Needle nose pliers are possibly the most used tool on the electronics workbench. They allow you to do things that your big, fat fingers just can't.
- Diagonal, or flush, cutters. You use diagonal cutters to cut wire and trim soldered leads.
- Wire strippers. A good pair of wire strippers is essential when making cables or when you have to solder wires to circuit boards.
- Terminal crimper. You use the crimper to properly attach terminals to wires. Make sure to also purchase

a selection of crimp-on terminals.

- Precision (jeweler's) screwdrivers. Many of the screws you'll find in electronics equipment are just too small to use normal-sized screwdrivers. A set of jeweler's screwdrivers will have a couple of Phillips-head screwdrivers as well as several conventional screwdrivers.
- Hobbyist knife. This is the type of knife that modellers use. It's just as handy in electronics work as it is in building models.
- Soldering iron or soldering station. Even if you're not going to be doing a lot of building, you need a soldering iron to make simple repairs and build simple cables. Being able to solder is an essential skill for a radio amateur.
- De-soldering tool. If you do any soldering, there will undoubtedly be times that you have to de-solder a connection. Buy a spring-loaded "solder sucker" and not a hand-operated desoldering bulb. The spring-loaded units work a lot better.

Other tools that you'll find useful if you intend to do a lot of building include:

- Anti-static mat and/or wrist strap. Many electronic components can be damaged by an electrostatic discharge. That's why you want to use an anti-static mat and/or wrist strap. These drain off static electricity so that you don't zap your electronics. Amazon, not surprisingly has a wide selection. You can also get them at Radio Shack.
- Tweezers. You need tweezers if you're working with very small components, such as surface-mount devices.
- Table vise. You need a table vise to hold a circuit board while your building or repairing it, or to hold a

connector that you're soldering wires onto.

- Lighted magnifier or magnifying visor. If you're north of 40 years old, then you need good lighting and probably some magnification. Some of the parts used today are very small, making the markings hard to read and making them difficult to handle. A magnifying light or magnifying visor makes working on circuits a lot easier.

If you're really starting from scratch, you might want to consider buying a complete tool kit. Sears (yes, Sears!) sells many different electronics tool kits. Some of the tool kits include a digital multimeter and soldering iron. The nice thing about buying a tool kit is that some kits include a carrying case. Other sources for toolkits include Jameco (www.jameco.com), Sparkfun (www.sparkfun.com), and the Electronic Tool Box (www.electronicstoolbox.com).

My own tool set has evolved over the years. I still have some needle-nosed pliers and some diagonal cutters that I acquired over 30 years ago when an electronics manufacturing company that I worked for took them out of service. I got a set of tweezers at some hamfest. The table vise I use is an el cheapo from Harbor Freight. You could do the same, acquiring the tools as you find them, but the problem with that is that they may not be on hand when you need them.

However you get your tools, make sure that you do get them or have access to them. If you can't make a cable or perform a simple repair because you don't have the tool to do it, it will be frustrating at the very least, and it could be expensive if you have to pay for a new cable or pay someone to make a repair for you.

Resources

- Collin's Lab: Electronics Tools (http://www.youtube.com/watch?v=Kv7Y8nAOoFE)

- How to equip your EE lab (http://www.ladyada.net/library/equipt/)

BUY A DIGITAL MULTIMETER (DMM)

A digital multimeter, or DMM for short, is the most basic piece of test equipment you can own, and every ham should have one. With a digital multimeter (DMM), you can make voltage, current, and resistance measurements. Some multimeters do even more, but that's a topic for another book.

Why do you need a multimeter? Well, the multimeter is the first thing you'll reach for when you have problems with your equipment. For example, let's say you go down to your shack, switch on your radio, and nothing. It doesn't turn on. The first thing you should check in this case is that the power supply is supplying the proper input voltage. To do this, you pull out your DMM, set it to measure voltage, place the probes on the + and - outputs, and verify that the power supply is working.

OK, now we're sure that the power supply is working OK, but the radio still doesn't power up. The next thing to check is the power cable from the supply to the radio. It's possible that the cable has an open connection. To check that, you first disconnect the cable from the power supply and from the radio.

Then, set your DMM to measure resistance. Set it on the

lowest resistance scale. Connect one test probe to one end of the cable and the other test probe to the other end. The resistance you measure should be very low—less than 2 or 3 ohms. An open connection will register an infinite resistance.

I think you get the picture. Without a DMM, you're dead in the water. With a DMM, you can figure out what's wrong and fix it.

This multimeter costs about $50 and features a rugged case that helps prevent damage should you accidentally drop it.

How to buy a DMM

There are a wide range of DMMs available. On the low end, you'll find DMMs at Harbor Freight for $5 or less. On the high end, you could spend $300 or more for a Fluke multimeter. I would advise against both. The $5 multimeters are not very well-made and can be inaccurate. They tend to quit working just when you need them.

The $300 DMMs are great, but you needn't spend that much. A DMM costing between $30 and $100 will do pretty much all you need to do at this point in your amateur radio adventure, and you can use the money you have left over for other things. You can buy them at any Lowe's or Home Depot. Ask your friends or Elmer what kind of meter they own and whether or not they would recommend that you buy something similar.

The DMM is a versatile instrument. Learning how to use one will help you troubleshoot problems quickly and make amateur radio a much more enjoyable hobby.

BUILD AN ANTENNA

Building an antenna is something that you should do soon after getting your license. There are many reasons for this, including:

- Building an antenna will help you learn how antennas really work.
- Building an antenna is cheaper than buying them.
- If you're using a handheld with the standard "rubber ducky" antenna, you can build an antenna that will increase the range of your handheld.
- It's fun!

Building a 2m quarter-wave ground-plane antenna

The first antenna that you should consider building is the quarter-wave ground-plane antenna for the 2m band. They are very easy to build and will perform better than the antennas that come with most handhelds.

The quarter-wavelength, ground plane antenna is made up of one vertical element, called the driven element, and four radials. The radials make up the ground plane. An easy way to make this antenna is to use an SO-239 coax connector. The driven element is soldered directly to the center conductor, while the four radials are connected to the four holes in the connector's flange. See the figure below.

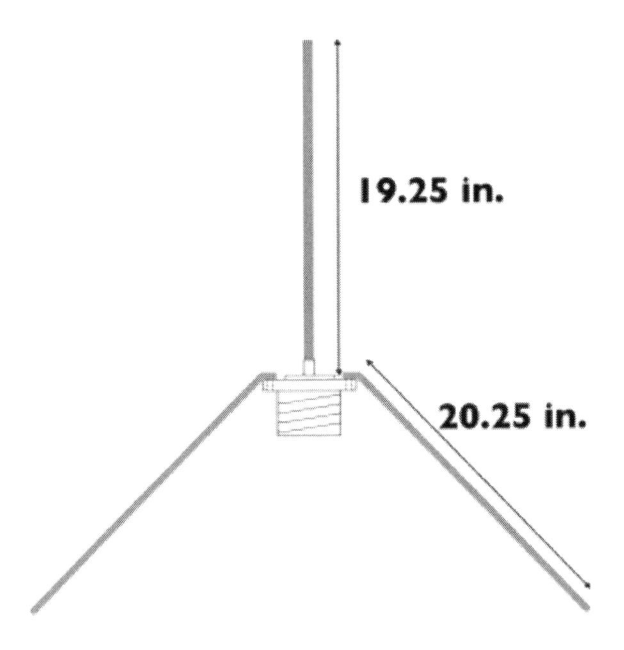

Now, let's calculate how long the elements should be. Since the wavelength of a radio wave is equal to 300/f (MHz), one quarter wavelength will be equal to 75/f (MHz). At 146 MHz, therefore, the length of the driven element is:

75/146 = .51 m

In practice, we have to make one more adjustment. Because a radio wave travels more slowly in a wire than it does in free space, the wavelength will actually be about 5% less in a wire than in free space. So, we multiply the wavelength in free space by .95 to get the length of the driven element:

.51m x .95 = .49m = 19.25 inches

The radials should be about 5% longer than the driven element. This isn't really very critical, so if you make them

20.25 inches long, the antenna will work just fine.

You should make the elements out of a stiff wire. 12 AWG copper wire will work for experimentation purposes. Welding rod might be better for a more permanent antenna.

You need to solder the 19.25-in. driven element to the solder cup of the center conductor of the SO-239 connector. Attach the radials to the holes in the flange of the SO-239 connector with nuts and bolts. You can also use these nuts and bolts to mount the antenna to some kind of bracket. Bend the radials out to a 45-degree angle, connect a coax cable to it, and start having fun!

For more information

For more information on how to build and what you can do with the quarter-wavelength, ground-plane antenna:

- Capital City Amateur Radio Club SO-239 antenna (http://www.w7tck.org/_misc/so-239_ant.html)
- Simple 2 Meter Ground Plane Antenna Project with PVC Support (http://www.hamuniverse.com/kc0ynr2metergppvc.html)

BUILD A KIT

While most amateur radio operators today buy their equipment rather than building it, a well-rounded amateur radio operator should have basic electronics construction skills. This includes knowing how to read a schematic diagram, being able to identify the different types of electronic components, and how to solder.

Building a kit is a good way to acquire these skills. Building a simple kit will teach you all of these skills, and once you've successfully completed the kit, you may even have something that's useful.

The PicoKeyer Plus makes a good first kit. It has less than 20 components, and once complete is a useful addition to your shack.

What kit should you build?

Kits are available from many different companies. Really too many to list here. Googling "electronic kit" turns up more than 12 million results!

What I can do here is to tell you about a couple of the kits we've built during our club construction nights. Each time we've done this, we have had 20 or more builders, and by the time it was ready to go home, everyone of them had his or her kit working. Usually, there are some people who've never even soldered before, but that didn't stop them from successfully completing their kit.

The first kit we built was the N0XAS PicoKeyer. N0XAS no longer produces this particular kit, but he's replaced it with one that's even better - the PicoKeyer Plus (www.hamgadgets.com). The reason that I chose this kit is that it is inexpensive (less that $20), has fewer than 15 components, and a very good manual that includes step-by-step assembly instructions.

A keyer is a device that is used to key a transmitter when operating Morse Code. The PicoKeyer allows you to set the speed at which you send code and has memories that allow you to automatically send frequently sent messages. If you're just learning Morse Code, you can use the PicoKeyer as a code practice oscillator.

Another kit that we built is the Sure PS-LP11111 5~16 VDC Linear DC Voltage Power Supply. This kit can be purchased from Amazon (http://www.amazon.com/5-16-Linear-Voltage-Power-Supply/dp/B005FMTCWA) for about ten bucks. This kit has less than 20 components, and when you're done with it, you can use with wall wart transformer to supply DC voltages for other projects. The downside to building this particular kit as your first construction project is that the instructions are very sparse. If you decide to build this kit, be sure to have someone who can help you should you have any trouble with it.

Building your kit

Here's what the PicoKeyer manual has to say about building their kit:

With just a little care and practice, even a first time kit builder can complete the project in a relatively short time. You will need to gather a few tools and supplies together before beginning to assemble your kit. Here's what you will need:

- A clean, level, static-free work area with good lighting. Wooden workbenches are fine. If you are working on a kitchen table, be sure to spread out some newspaper or something else to keep solder splatters and sharp wire ends from damaging the table top.
- A soldering iron. A small, low-wattage (25-35 Watt) pencil type iron is ideal. Avoid larger, pistol-grip types. You can find inexpensive irons at your local Radio Shack. You will need a fine tip intended for electronics. Be sure to use an iron rest or holder to keep the iron from damaging your work surface. If you plan to assemble more kits, I recommend investing in a good quality, temperature controlled soldering station such as the Weller WES or WLC series. You'll be glad you did! Follow the iron manufacturer's instructions for tinning the tip, and keep a damp sponge handy to keep the tip clean.
- Solder suitable for electronics work. Use a good quality, small diameter rosin core solder intended for electronic assembly. DO NOT use acid core solder!
- Small needle-nose pliers and a pair of small diagonal wire cutters. The smaller you have, the better off you will be. Again, you can find hand tools intended for electronics work at Radio Shack and other suppliers such as Techni-Tool, Jensen, Mouser and Sears.
- A clamp or small vise to hold the work is a good idea. I use a PanaVise, but you can also construct a board

holder out of scrap wood and rubber bands. If you use a regular bench vise, use gentle pressure and something to cushion the vise jaws.

- A pencil to check off each step as you finish it.

You can do it

You really can do this, and the skills you learn will make you a better amateur radio operator. Not only that you'll be surprised at how much fun building your own gear can be. At our club's first build night, we had a young woman who was building her first kit. I will always remember her squeal of delight when we inserted the battery and her keyer came to life. There are very few things like that feeling.

GO TO A HAMFEST

You can often buy stuff like power cords and connectors at bargain prices at a hamfest.

When I was a kid in Michigan, we used to call a ham radio swap meet a "swap and shop." Nowadays, they're mostly known by the term "hamfest." Whatever name you know them by, they're both educational and a lot of fun.

There are a lot of reasons to go to a hamfest, including:

- You get to see a lot of ham radio gear in one place.
- You might be able to get a good deal on some used (or new) equipment.

- You might find something that will be fun to play with.
- You get to meet hams face-to-face that you've only talked to on the air.

You never know what you'll find at a hamfest. If it's a decent-sized hamfest, chances are you'll find equipment ranging from radios made in the 1950s with vacuum tubes to modern computer-controlled transceivers. If nothing else, you'll get an education on the wide range of amateur radio equipment that's out there.

Can you get a good deal on a radio? Possibly, although these days so much stuff is sold on EBay and via the online ham classifieds on QRZ.Com, eHam.Net, and other sites, that getting a real "steal" is getting harder and harder. One thing is for sure, if you're a new ham and don't really know how to evaluate a particular piece of equipment, get your Elmer to look over a purchase before you hand over your money. What may look like a bargain, may end up costing more than a new radio.

What you can often get a good deal on are small parts, such as connectors, power cords, speakers, etc. You never know when you'll need a 1/4-in. phone plug to put on the end of a set of headphones. A friend of mine jokes that at every hamfest he always buys a handful of different connectors. Hamfests are good places to stock up on these types of things.

You'll find more than used equipment at a hamfest, though. Many dealers will bring new equipment to a hamfest, especially if it's one of the big hamfests. This is your chance to look at a number of different radios that you may have only been able to look at in catalogs and compare different models. In addition, dealers often offer "hamfest prices," so you may be able to get that radio at a slight discount.

Hamfests are also good places to connect with other hams.

Quite often, you'll meet guys that you've only talked to on the air. It's a lot of fun to connect a name and callsign with a face. Sometimes, different ham groups, such as ARES/RACES groups or QRP clubs, will set up a table to promote their group. You can use this opportunity to find out more about these groups and their activities.

To find a hamfest near you, go to the ARRL Hamfests and Conventions Calendar page (http://www.arrl.org/hamfests-and-conventions-calendar).

LEARN THE LINGO

Like many subcultures, amateur radio has a lingo all its own. Tune into a conversation on 40m phone, and you're likely to hear something like this:

> "W8ABC, this is K0XYZ. You're 5 by 9 here, but there's a lot of QR-Mary."
>
> "K0XYZ, this is W8ABC. QSL, Fred. Thanks for the report. I'll say 73 on this one."
>
> "Fine business, Joe. Catch you later. K0XYZ clear."

The first thing you'll notice are the Q-signals. Q-signals are three-letter codes that are used mostly when communicating with Morse Code, but their use has become common in voice operation as well. They were originally developed for use by maritime radio operators, but they were also adopted by airborne radio operators, and, of course, by amateur radio operators.

When operating Morse Code, a Q-code takes the place of an entire phrase. So, for example, if I send "QTH ANN ARBOR MI," what I mean is, "My location is Ann Arbor,

Michigan." Appending a question mark, makes the Q-code a question. "QTH?" would mean, "What's your location."

A complete list of Q-signals used in amateur radio can be found at http://www.qsl.net/w5www/qcode.html. Many of these are not frequently used, though, even when operating Morse Code. In addition to QTH mentioned above, some of the most commonly used Q-signals are:

- QRL - I am busy or the frequency is busy. One sends QRL? before calling CQ to determine if a particular frequency is in use.
- QRM - You are being interfered with.
- QRN - I am receiving a lot of atmospherice noise.
- QRP - Lower power. QRP is often used as an adjective. A low-power transceiver, for example, might be called a "QRP rig."
- QRS - Send slower.
- QRT - Stop sending. QRT is often used as a verb. "I am going to QRT" means that you plan to go off the air.
- QRZ? - This Q-signal is almost always used with a question mark. It means who is calling me?
- QSB - Your signal is fading in and out. QSB, QRN, and QRM are often used as nouns to mean fading, noise, and interference, respectively.
- QSO - I can communicate directly with [a particular station]. This Q-signal is also often used as a noun. "I had a QSO with Joe on 40m last night" means that I contacted Joe on the 40m band last night.

Hams also use phonetic alphabets when operating voice communications. In the example above, K0XYZ notes that there is a lot of "QR-Mary," Mary being the phonetic way to say the letter "M." We use phonetics because many letters sound alike, especially over a noisy radio channel.

The ARRL recommends that amateurs use the NATO phonetic alphabet (http://www.wikipedia.org/wiki/

NATO_phonetic_alphabet) as most amateurs around the world will recognize them. It was developed in the 1950s, and was designed to be intelligible and pronounceable by all the NATO allies.

Finally, a lot of the lingo we use in amateur radio is a holdover from the days when all amateur radio communication was in Morse Code. An interesting example of this is the use of the term "fine business," or FB in Morse Code.

Fine business can mean "OK" as in, "FB, Joe. I copied all of that last transmission." It can also mean "good" or "excellent." In Morse Code, one might send "THE KX-1 IS A FB RIG JOE."

Finally, when amateur radio operators end a contact, they often say "seventy three" or "seven three." In this context, "seventy three" means "best regards." The origin of this term is as old as amateur radio.

Before the Internet made long-distance communications so cheap, amateur radio was often used to send messages across the U.S. and around the world. Many common messages were codified to make sending them quicker. For example, if you wanted to wish your Aunt Harriet in Poughkeepsie a happy birthday, you'd get hold of an amateur radio operator. Instead of sending the text, "Greetings on your birthday and best wishes for many more to come," he would simply send "FORTY SIX."

SEVENTY THREE is short for "best regards." So, when we sign off with that number, we're wishing the ham on the other end of the QSO a fond farewell. If you're particularly fond of the ham on the other end, you might say "EIGHTY EIGHT." Be careful, though, "EIGHTY EIGHT" is short for "love and kisses."

Finally, there's some lingo that you might want to *un-learn*. By that I mean the lingo used on the Citizen's Band. We don't use 10-signals in amateur radio, and doing so will not win you friends in the amateur radio community. So, if you come from the CB world, keep that in mind when you get on amateur radio.

SUBSCRIBE TO MAILING LISTS, BLOGS, AND PODCASTS

When you're just starting out in amateur radio, you want to learn as much as you can about the hobby. One way to do this is to find an Elmer (see chapter 1). In this age of the Internet, another great way to do this is to join ham radio mailing lists and subscribe to ham radio podcasts. These resources give you access to hundreds, if not thousands, of Elmers.

One mailing list that I always suggest to new hams is the HamRadioHelpGroup (groups.yahoo.com/group/HamRadioHelpGroup). The purpose of this group is to help "those who are interested in getting started in Amateur Radio or upgrading their license." This mailing list has a good mix of beginners and experts, and most questions are answered quickly and correctly. One thing that I really like about this group is that the moderators do a good job of keeping the discussions on track, and will squelch them when they stray off topic or threaten to turn into flame wars.

In addition to the HamRadioHelpGroup, you might also want to join a more targeted mailing list. For example, if you're interested in learning Morse Code (hint, hint), you might join the SolidCpyCW list (groups.yahoo.com/

group/SolidCpyCW/). If you just bought a Yaesu FT-60 hand-held transceiver, you might want to join the FT-60 list (groups.yahoo.com/group/FT-60/). Chances are that no matter what your interest, there's probably a mailing list to discuss that interest.

I'm subscribed to a lot of amateur radio mailing lists and could probably spend most of my day just reading and replying to them. In order to get the most out of them, without them taking away from my on-air time, I only read those threads that I am really interested in, and even then, I quit reading them once they have started to drift off-topic. I also un-subscribe myself from lists that cover topics that I'm no longer interested in.

Blogs, podcasts and videos

In addition to getting on a few mailing lists, you might want to read a few blogs and subscribe to podcasts. These are also great sources of information about amateur radio. I blog about amateur radio at www.kb6nu.com, and lots of hams find it a good source of information. You can find a list of other ham radio blogs that I'd recommend on my home page.

Podcasts are also a good source of information. One podcast that you might want to check out is the Practical Amateur Radio Podcast (www.myamateurradio.com). Since 2008, Jerry, KD0BIK, has been producing PARP, and currently has more than 50 different episodes online. For other podcasts, consult the list on Jerry's home page.

Finally, there are literally thousands of amateur radio videos on the net. On YouTube alone, there are approximately 32, 000 of them. The American Radio Relay League has its own channel (www.youtube.com/user/ARRLHQ), but perhaps the most popular amateur radio video channel is the K7AGE channel (www.youtube.com/user/K7AGE). K7AGE has more than 6,200 subscribers and his videos have garnered more than 2.1 million views!

Whatever source or sources of information you select,

remember to not let them take up too much of your time. Ham radio is about more than just reading, listening, or watching. It's about doing!

UPGRADE TO GENERAL

As soon as you pass the Tech test, you should start studying for your General Class license. While there are certainly many fun and useful things you can do as a Technician, there are several reasons that you will have more fun as a General Class amateur radio operator.

One reason to upgrade right away is that you're in study mode already. There's no need to get back in the habit of studying and getting used to taking tests again. While the General Class test is more difficult than the Tech test, there are lots of good resources, both in print and on the Internet, to help you pass. These include my *No-Nonsense General Class License Study Guide* (http://www.kb6nu.com/study-guides/).

Another reason upgrade to General Class is that it will make you a better ham. Seriously. Even if you memorize the answers to all of the questions in the question pool, you're bound to learn something. That knowledge could come in handy when you want to put up a new antenna, buy a new radio, or determine the best band to use for a particular communication.

Perhaps the best reason to upgrade to General is that it gives you more privileges on the HF bands. While talking on the local repeater may be fun, let's face it, there are only so many folks to talk to. By getting on the HF bands, you'll be able to talk to thousands of hams all around the world, not just around the corner. That's why many hams, including me, think that the shortwave (HF) bands are where the magic happens. As a General Class operator, you'll get access to all the HF bands, including 20m, and you get to operate phone on them as well as CW.

Operating the HF bands literally expands your horizons. You get to meet hams from all over the world, not just around the corner. If you haven't yet operated on the HF bands, you don't know what you're missing. To get a taste of HF operation, ask a ham with an HF station if you can visit him sometime and watch him operate. Chances are he'll let you make a few contacts of your own. You may also get the opportunity to operate an HF station during Field Day or if your club has its own club station.

Another thing that you can do as a General Class licensee is become a Volunteer Examiner (VE). Becoming a Volunteer Examiner and helping others get involved in amateur radio is a great way of giving back.

A General Class license will let you do more and have more fun with amateur radio. If you haven't yet taken the Technician test, you should be aware that you can take both the Tech and the General (and even the Extra) Class tests at the same test session. If you prepare for both tests, you could walk out of your first test session with a General Class license and skip the Technician Class altogether.

GO TO FIELD DAY

Field Day (http://www.arrl.org/field-day), held on the last full weekend in June, is the quintessential amateur radio event. It includes elements of just about everything that makes amateur radio the great hobby that it is, and you should make every effort to participate in Field Day the first year that you're licensed.

Tents often serve as shelters for Field Day stations.
Photo courtesy of Ken Barber, W2DTC.

Field Day got its start in 1933 as an emergency-communication exercise. Ham radio operators dragged their equipment out into a field somewhere and operated using emergency power sources. The aim was to see how prepared amateur radio operators were to respond to an emergency and to learn how to do it better.

Emergency communications preparedness is still the primary purpose of Field Day. Amateur radio operators tune up their gasoline-powered generators and test their solar panels to ensure that they will be ready in case of an emergency. And, by hauling out into the field all manner of radio equipment, we find out what radios will work best in that operating environment.

Of course, the only way to tell how well your equipment will work is to actually operate it. That's where the contest part of Field Day comes in. Stations score points by making contacts with other stations, and those with the most points win. Other things being equal, the stations that work the best will make the most contacts and score high in the contest.

Many Field Day stations have multiple transmitters, and when you have multiple transmitters, you need multiple antennas. Setting up a multiple-transmitter operation can be a lot of work. That's why Field Day is often a club activity. For some clubs, it's the biggest event of the year. In addition to all the technical activities, clubs use Field Day as a social event. There's food and drink and reminiscing about Field Days gone by. For some hams, that's more fun than actually operating.

Finally, because Field Day is such a big event, the ARRL encourages us all to use the event to reach out to the public, elected officials, and served agencies, such as county emergency management and the Red Cross, and educate them about amateur radio. Unlike many contests, where you only score points when you make contacts, you score Field Day points for holding your operation in a public place, handing out brochures to interested parties, and having the mayor come and visit your Field Day site.

How to participate

By participating in Field Day, you'll learn more about amateur radio in a single day than you will doing just about anything else. If you're a club member, ask how you can help out organizing your club's Field Day event. That's sure to win you points, and it will make your Field Day experience that much more fun and educational.

If you're not a club member, or if you'll be out of town that particular weekend, you can find a Field Day site closeby, by going to http://arrl.org/field-day-locator. The clubs that are listed there are sure to welcome you, especially if you arrive early and help them set up.

I hope I've persuaded you to participate in the next Field Day. You'll not only learn a lot, but you'll have a lot of fun. Don't forget to take some sun screen and mosquito repellent!

LEARN MORSE CODE

Before you even start reading this chapter, I'll warn you that I'm a big fan of Morse Code (often referred to as CW, or "continuous wave"). So big, in fact, that it's safe to say that I use Morse Code to make 95% of my contacts.

I am not, however, one of those guys that thinks you're not a "real ham" if you didn't pass some kind of code test. In fact, I think that eliminating the code test was a good thing for ham radio. The code test kept a lot of good people out of the hobby.

Having said that, I think there are lots of good reasons you should learn Morse Code. Please keep an open mind as I list them:

- Tradition. Operating CW is an amateur radio tradition. When amateur radio began, CW was the only mode. When you learn and operate CW, you're following a very long line of hams who have operated CW.
- Effectiveness. Talk to a CW operator, and it's likely that he'll chew your ear off about how CW is a more effective mode than voice. While the difference is probably not as much as many CW operators would like you to believe, the difference is real. When

conditions are poor, you'll be able to make CW contacts and not voice contacts.

- DXing. That being the case, CW operators have an advantage when it comes to contacting DX stations because their signals will get through when voice signals are unreadable. Also, if you consider that there are more voice operators than CW operators, you'll have a better chance of contacting a much-wanted DX station because there will be fewer operators trying to contact him using CW than there will be using voice.
- Contesting. In most contests, you get more points for a CW contact than you do for a voice contact. Sometimes the bonus is 100%, sometimes only 50%. In either case, doesn't it make sense to know CW if you want to be a contester? You'll score more points for the same number of contacts.
- Simplicity/Efficiency. The equipment you need to operate CW is a lot simpler than the equipment needed to operate voice modes. And, because CW is more efficient, you can, in general, use a lot less power to make contacts with CW than you need to make contacts using voice modes. This has spawned a whole sub-group of hams called QRPers, who delight in using very minimal equipment to make contacts.
- Using CW also saves bandwidth. The bandwidth of a CW signal is approximately 200 Hz, while the bandwidth of a single-sideband (SSB) voice signal is about 3 kHz. That is to say that the voice signal is 15 times wider than the CW signal. Another way to say this is that for a given amount of bandwidth, you can fit 15 times more CW signals than you can SSB signals.
- It's just plain fun. Once you learn CW and start using it, it can be a lot of fun. Like any activity that requires some skill, mastering that skill can be a source of

pride. Not to sound too vain about it, but I enjoy the praise I get from my fellow hams when I can display my CW operating skills.

How to Learn Morse Code

In the old days if you wanted to learn Morse Code, you went out and bought a vinyl record or maybe a cassette tape that had precrecorded lessons on them. Another approach—the approach I used—was to tune in a Morse Code signal and start to associate the patterns of dits and dahs to characters of the alphabet. Both methods had drawbacks.

Today, things are a lot easier. Not only are there free resources available, I think they are much more effective in teaching people code than the old LPs or cassette tapes. Here are the three resources that I recommend:

- G4FON Koch CW Trainer (www.g4fon.net). Ray Goff, G4FON, has perhaps written the most popular CW training program. It runs on the PC, and is completely free! The program uses the Koch method. The idea is that you learn to receive at the speed you would like to eventually achieve, but you learn only one character at a time. This method works very well for lots of people.
- K7QO Code Course. The K7QO Code Course takes a different approach. This set of .mp3 files comes on a CD-ROM and teaches you the code letter by letter. It starts out sending the letters slowly, then ramps up. The nice thing about this course is that you can use it on any device that is capable of playing .mp3 files. To obtain a copy of the CD-ROM, send $1 per copy and a self-addressed envelope to FISTS, PO Box 47, Hadley MI 48440.
- Learn CW Online (lcwo.net). LCWO uses the Koch method to teach Morse Code. Because it runs in your browser, you can use this website no matter what computer you happen to be using.

Whatever method you choose, I hope you'll consider learning the code. See you on the CW bands!

GET TO KNOW YOUR (HAM) NEIGHBORS

Jukka, OH2BR, gave me this bit of advice. "Find out who your closest ham neighbors are," he said, "and contact them." "They could be your best friends and Elmers OR your worst opposition if you interference to their ham activities. Start early—contact them today!" He went on to tell this story:

> "I found out how important it is to acquaint myself with my neighbor hams the hard way. I built a 15W XTAL TX and started working the world. One day I got a note from the President of our national league SRAL (Finland), OH2TK. He lived one block away from me. Was I terrified! I thought I would be expelled from the League for causing interference with my poorly constructed TX. Happily, he was a most amiable person and took me under his wing, so there was a happy end to the story."

Today, there are many ways to find the hams in your neighborhood. One way to do this is to do it the old-fashioned way—walk around your neighborhood looking for antennas.

Another way to do this is to visit QRZ.Com. Most of us use QRZ.Com to search for particular amateur by typing his or her callsign into the search box. You can, however, get a list of ham radio operators in your zip code by typing that into the search box. When I typed "48103" into the search box, it returned 150 licensees.

Perhaps an even better way to do this is with Where are All the Hams? (http://hams.mapmash.com/hammap.php). When you type you zip code into the appropriate box on this Web page, you get a Google map that shows where the hams are located. You can then zoom in and pan around to find the hams closest to you.

So, find the hams in your neighborhood and get to know them. You never know. You might make a friend for life.

BUY QSL CARDS

Once you start making contacts, other amateurs will want to swap QSL cards with you, even if you just talk to them on the local repeater. The purpose of a QSL card is to confirm that you had a contact with another amateur. For sure, you'll want to have some cards printed up if you operate on the shortwave bands. Sometimes, amateur radio operators call swapping QSL cards "the final courtesy."

Once you get started swapping QSL cards, you may get hooked on QSLing, and it certainly can be an enjoyable part of the hobby. Many designs are distinctive, and they are fun to show off to friends and family. When I speak to groups about amateur radio, I always bring a selection of QSL cards that I've received. They can be very impressive.

Another reason to collect QSL cards is that they're often needed to qualify for awards and certificates. You can, for example, get the Worked All States Award from the ARRL by submitting a QSL card from a station that you contacted in each of the 50 states.

Collecting QSLs can be fun, even if you don't plan to apply for an award. I have, for example, started a small collection of QSL cards from stations whose callsigns spell words. I now have more than 150 such QSL cards including cards from W8HOG, WB4DAD, N4HAY, and KD8EGG. I agree that it's kind of odd, but it's fun, too.

Where to get QSL cards

There are many companies that print QSL cards. Here are some in no particular order, and with no endorsement implied:

- CheapQSLs.Com
- KB3IFH QSL Cards (http://kb3ifh.homestead.com/)
- UX5UO world of QSLs (http://www.ux5uoqsl.com/)
- QSLs by W4MPY (http://www.w4mpy.com)

All of these companies offer stock designs, but can also print custom designs. I suggest starting out with one of the stock designs and then consider a custom design once you've run out of the first printing. Below is the card that we use for our club station, WA2HOM, at the Ann Arbor Hands-On Museum.

JOIN SKYWARN, ARES, OR RACES

One of the principles upon which the amateur radio service is founded is that, when needed, amateur radio operators will provide public service and emergency communications. Part 97.1 (a) reads:

> "Recognition and enhancement of the value of the amateur service to the public as a voluntary noncommercial communication service, particularly with respect to providing emergency communications."

Think of it as part of the price you pay for the privilege of being granted an amateur radio license.

One way to get involved with public service and emergency communications is to join SkyWarn (http://www.skywarn.org). SkyWarn is a volunteer program run by the National Weather Service with more than 290,000 trained severe weather spotters. These volunteers help keep their local communities safe by providing timely and accurate reports of severe weather to the National Weather Service.

Not all of these weather spotters are amateur radio operators, but a good many are, and using amateur radio is a great way to report severe weather. When severe weather is imminent, SkyWarn spotters are deployed in the areas where the severe weather is expected. A "net" is established on one of the local repeaters, and all of the SkyWarn spotters who have amateur radio licenses check into that net. The net control advises the spotters when they might expect to see severe weather, and the spotters, in turn, report conditions such as horizontal winds, large hail, rotating clouds, and even tornadoes.

To become a SkyWarn spotter, you must take a class that teaches you the basics of severe weather, how to identify potential severe weather features, and how to report it. The classes are free and typically last about two hours.

ARES/RACES

Another way to become involved in public service and emergency communications is to join an ARES/RACES group. Although technically these are two separate services—the Amateur Radio Emergency Service (ARES) is run by the ARRL, while Radio Amateur Civil Emergency Service (http://www.usraces.org/) is a function of the Federal Emergency Management Agency (FEMA) —the amateur radio operators who typically take part in one also take part in the other.

To participate in RACES, you'll need to to take some self-study FEMA course in emergency preparedness and emergency-response protocols. Classes may or may not be required to participate in ARES. These requirements are set by each individual ARES group. To get involved with either ARES or RACES, ask your local club members when they meet. You can also contact the Section Manager or Emergency Coordinator for your ARRL section. To get in touch with those people, go to http://www.arrl.org/sections and find the section that you live in.

If these formal organizations aren't for you, you can still participate in public service activities through your club. Our club, for example, provides communications for a bike tour with more than 1,000 riders and covering dozens of square miles. Our organization is a lot less formal than SkyWarn, ARES, or RACES, but the public service that we provide is just as valuable.

PARTICIPATE IN A CONTEST

Human beings are competitive by nature, and since amateur radio operators are human, they find ways to compete with one another. Almost every weekend—and some weekdays, too—there's some kind of amateur radio contest. They are a lot of fun, and all classes of amateur radio operators can participate.

Most contests have some kind of theme. For example, nearly every state has what's called a QSO party. During a state's QSO party, stations outside the state get points for contacting as many stations in as many counties inside the state, while stations in the state get points for contacting stations outside the state as well as inside the state. There are also QRP contests, where all stations must operate with low power and DX contests, where the goal is to work stations outside your own country.

Most of these contests take place on the HF bands, but even as a Technician you can participate in these contests if you know Morse Code. If you haven't yet cracked the code, you can still participate in the contests that take place on the 10m band and above. Another way to participate is to be one of the operators in a multi-operator setup. As long as one of the operators with a General Class or Extra Class license acts as the control operator, you can operate in those portions of the bands where you don't have privileges.

I prefer operating in the smaller contests, such as the state QSO parties, to operating in the big contests, such as the CQ Worldwide DX contest or the ARRL Sweepstakes. There are a lot fewer stations competing and the bands are a lot less crowded. Sometimes with even a modest effort, you can earn an award. It's also easier to compete in a smaller contest with a modest station—like the one I have—than it is to compete with the big guns in the major contests.

One way to get started might with the ARRL's Rookie Roundup. This contest was designed to get newcomers involved in contesting. It takes place three times per year in April, August, and December, and lasts for six hours. Rookies score points for all their contacts, while "old timers" only score points by contacting "rookies."

I hope you'll give contesting a try. They're a lot of fun and a big part of the amateur radio hobby.

Resources

- National Contesting Journal (http://www.ncjweb.com/). The National Contest Journal is published six times per year (Jan/Feb, Mar/Apr, May/Jun, Jul/Aug, Sep/Oct and Nov/Dec) and is dedicated to covering the competitive contesting aspects of amateur radio. Each issue is loaded with information of interest to contesters (and DXers, too!); from casual observer to hardcore competitor, from little pistol to big gun.

- WA7BNM Contest Calendar (http://www.hornucopia.com/contestcal/). This site provides detailed information about amateur radio contests throughout the world, including their scheduled dates/times, rules summaries, log submission information and links to the contests' official rules.

HAVE FUN!

I've given you a lot of advice in this book so far. I've suggested that you find an Elmer, join a club, learn Morse Code, buy QSL cards, etc., etc., etc. This chapter has the best bit of advice, though—just have fun..

If you're not having fun with your club, quit. If you don't like Morse Code, don't do it. Amateur radio is a hobby, and if you're not having fun doing some of these activities, then don't do them. What you should do is find those activities that you do enjoy doing and do those instead.

Amateur radio has been a great hobby for me, and I have had a lot of fun with it. I hope that it will be as much fun for you as it has been for me.

ABOUT THE AUTHOR

I have been a ham radio operator since 1971 and a radio enthusiast as long as I can remember. In addition to being an active CW operator on the HF bands:

- I blog about amateur radio at KB6NU.Com.
- I am the author of the "No-Nonsense" license study guides. See http://www.kb6nu.com/study-guides for more information.
- I send out a monthly column to more than 300 amateur radio clubs in North America for publication in their newsletters.
- I am the station manager for WA2HOM (www.wa2hom.org), the amateur radio station at the Ann Arbor Hands-On Museum (www.aahom.org).
- I teach amateur radio classes around the state of Michigan.
- I serve as the ARRL Michigan Section Training Manager and conduct amateur radio leadership workshops for amateur radio club leaders in Michigan.

You can contact me by sending e-mail to cwgeek@kb6nu.com. If you have comments or question about any of the stuff in this book, I hope you will do so.

73!

Dan, KB6NU

26469140R00045

Made in the USA
Middletown, DE
30 November 2015